THE TURING TESTS

KU-187-254

EXPERT
NUMBER
PUZZLES

Foreword by Sir Dermot Turing

This edition published in 2021 by Arcturus Publishing Limited
26/27 Bickels Yard, 151–153 Bermondsey Street,
London SE1 3HA

Copyright © Arcturus Holdings Limited
The Turing Trust logo © The Turing Trust

All rights reserved. No part of this publication may be reproduced,
stored in a retrieval system, or transmitted, in any form or by any means,
electronic, mechanical, photocopying, recording or otherwise, without
prior written permission in accordance with the provisions of the
Copyright Act 1956 (as amended). Any person or persons who do any
unauthorised act in relation to this publication may be liable to criminal
prosecution and civil claims for damages.

ISBN: 978-1-78888-753-3
AD006772NT

Printed in the UK

CONTENTS

FOREWORD

Alan Turing's last published paper was about puzzles. It was written for the popular science magazine *Penguin Science News*, and its theme is to explain to the general reader that while many mathematical problems will be solvable, it is not possible ahead of time to know whether any particular problem will be solvable or not.

Alan Turing's work at Bletchley Park is well known: untangling one of the most strategically important puzzles of World War II, the Enigma cipher machine. The Enigma machine used a different cipher for every letter in a message; the only way to decipher a message was to know how the machine had been set up at the start of encryption, and then to follow the mechanical process of the machine. The codebreakers had to find this out, and the answer was not in the back of the book. To begin with, they had squared paper and pencils, and they had to work out the cipher-machine's daily settings, using intuition and ingenuity. These characteristics constitute mathematical reasoning, according to Alan Turing, who was confident that there was no difference between the reasoning processes of a human provided with pencil, paper and rubber, and those of a computer.

Although they did not have computers to help them at Bletchley Park, with Alan Turing's help new electrical and electronic devices were invented which sifted out impossible and unlikely combinations and so reduced the puzzle to a manageable size. And the experience with these new machines laid the foundation for the development of electronic digital computers in the post-war years.

Computers are now commonplace, not only in the workplace and on a desk at home, in a smartphone or tablet, but in almost every piece of modern machinery. Teaching people computer skills and coding are now considered obvious elements of the curriculum. Except that this is not so in all parts of the world. In Africa, access to computers in schools is extremely variable, and in some countries there is little or no opportunity for students to have hands-on

experience of a real computer. For example, in Malawi students may have only a 3% chance of using a computer at school.

The Turing Trust, a charity founded by Alan Turing's great-nephew James in 2009, aims to confront these challenges in a practical way which honours Alan Turing's legacy in computer development. The Turing Trust provides quality used computers to African schools, enabling computer labs to be built in rural areas where students would otherwise be taught about computers with blackboard and chalk. The computers are refurbished and provided with an e-library of resources relevant to the local curriculum, and then sent out to give a new purpose and bring opportunity to underprivileged communities. The Turing Trust's projects in Malawi have since increased the number of secondary schools with computers in the Northern Region of Malawi from 3% to 44%. This has enabled thousands of students to start using these transformative technologies for the first time.

Thank you for buying this book and supporting the Turing Trust.

Sir Dermot Turing
February 2021
To find out more, visit www.turingtrust.co.uk

Notes to the reader
The puzzles in this book are not intended for the faint-hearted, but are designed to challenge experienced puzzle solvers. They are graded in three levels of difficulty, with the puzzles in the third level being truly for experts.

Unless otherwise stated the quotes in the book are by Alan Turing.

DOMINO PLACEMENT

A standard set of 28 dominoes has been laid out as shown. Can you draw in the edges of them all? The check-box is provided as an aid, and the domino already placed will help.

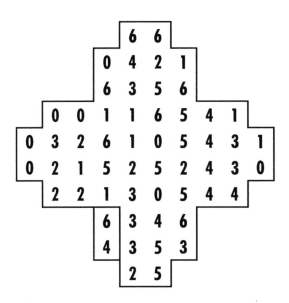

0-0	0-1	0-2	0-3	0-4	0-5	0-6	1-1	1-2	1-3	1-4	1-5	1-6	2-2

2-3	2-4	2-5	2-6	3-3	3-4	3-5	3-6	4-4	4-5	4-6	5-5	5-6	6-6
										✓			

ISOLATE

Draw walls to partition the grid into areas (some walls are already drawn in for you). Each area must contain two circles, area sizes must match those numbers shown next to the grid and each '+' must be linked to at least two walls.

2, 3, 3, 3, 3, 4, 7

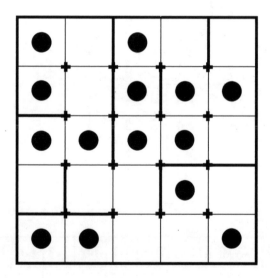

ROUND UP

Every circle contains a number which is the sum of
the numbers in the two circles below it.

Just work out the missing numbers!

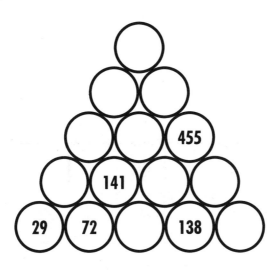

*'We can only see a short distance
ahead, but we can see plenty
there that needs to be done.'*

TOTAL CONCENTRATION

The blank squares below should be filled with whole numbers between 1 and 30 inclusive, any of which may occur more than once, or not at all.

The numbers in every horizontal row add up to the totals on the right, as do the two long diagonal lines; whilst those in every vertical column add up to the totals along the bottom.

								70

	16	2		21		5	114
6	14	17	20		1		86
23	12	2			15	30	106
22	8		18	17		8	112
2	4			21	25	18	106
5		26	3	12	14		98
	22	21	1	17		9	104
102	86	102	100	110	109	117	106

HEXAGONY

Can you place the hexagons into the grid, so that where any hexagon touches another along a straight line, the number in both triangles is the same?

No rotation of any hexagon is allowed!

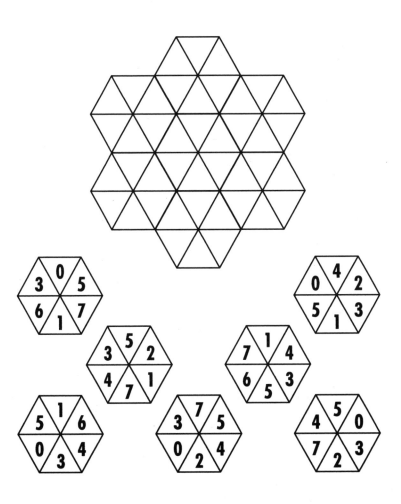

11

THE BOTTOM LINE

Can you fill each square in the bottom line with the correct digit?

Every square in the solution contains only one digit from each of the lines above, although two or more squares in the solution may contain the same digit.

At the end of every row is a score, which shows:

a the number of digits placed in the correct finishing position on the bottom line, as indicated by a tick; and

b the number of digits which appear on the bottom line, but in a different position, as indicated by a cross.

SCORE

4	5	2	6	✓
5	8	2	4	✓✓
2	2	7	4	✓✓
6	7	7	4	✓✓
3	4	1	5	✗
				✓✓✓✓

COIN COLLECTING

In this puzzle, an amateur coin collector has been out with his metal detector, searching for booty. He didn't have time to dig up all the coins he found, so has made a grid map, showing their locations, in the hope that if he loses the map, at least no-one else will understand it...

Those squares containing numbers are empty, but where a number appears in a square, it indicates how many coins are located in the squares (up to a maximum of eight) surrounding the numbered one, touching it at any corner or side. There is only one coin in any individual square.

Place a circle into every square containing a coin.

		1							
1	3		3			1	2	1	
2					2				1
3			4		3			4	
								4	
2		0		3					1
			3			0			
				2					
2						2	2	1	
	0			1	1				

LATIN SQUARE

The grid should be filled with numbers from 1 to 6, so that each number appears just once in every row and column. The clues refer to the digit totals in the squares, eg A 1 2 3 = 6 means that the numbers in squares A1, A2, and A3 add up to 6.

1 D E F 2 = 7

2 C D 3 = 7

3 B C 4 = 10

4 A B C 5 = 10

5 E F 6 = 4

6 A 1 2 = 7

7 B 1 2 = 5

8 C 1 2 = 9

9 D 4 5 = 6

10 E 4 5 = 9

11 F 3 4 = 9

	A	B	C	D	E	F
1						
2						
3						
4						
5						
6						

ZIGZAG

The object of this puzzle is to trace a single path
from the top left corner to the bottom right corner
of the grid, moving through all of the cells in either
a horizontal, vertical, or diagonal direction.

Every cell must be entered once only and your path should take
you through the numbers in the sequence 1-2-3-4-1-2-3-4, etc.

Can you find the way?

1	2	2	1	4	1	3	2
4	3	3	3	2	4	4	1
1	4	1	3	3	1	3	2
2	4	4	2	2	4	3	1
3	1	2	1	4	1	2	4
4	1	3	1	2	3	4	3
3	2	4	2	1	2	3	2
2	1	3	4	3	4	1	4

COMBIKU

Each horizontal row and vertical column should contain different shapes and different numbers.

Every square will contain one number and one shape, and no combination may be repeated anywhere else in the puzzle.

◇ ○ ☆ ⬡ □

1 **2** **3** **4** **5**

☆	⬡5	4		◇3
		☆2		
4				
○				
□				2

IT DOESN'T ADD UP

In the square below, change the positions of six numbers, one per horizontal row, vertical column, and long diagonal line of six smaller squares, in such a way that the numbers in each row, column, and long diagonal line total exactly 246.

Any number may appear more than once in a row, column, or line.

39	13	24	25	68	39
38	44	66	15	41	45
41	74	41	23	41	46
33	49	41	59	20	59
36	43	21	58	39	45
63	26	49	28	57	27

MIND OVER MATTER

Given that the letters are valued 1–26 according to their places in the alphabet, can you crack the mystery code to reveal the missing letter?

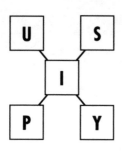

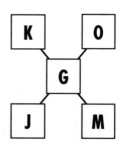

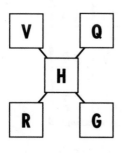

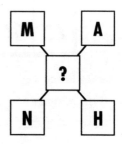

SUM CIRCLE

Fill the three empty circles with the symbols +, − and x in some order, to make a sum which totals the number in the middle. Each symbol must be used once and calculations are made in the direction of travel (clockwise).

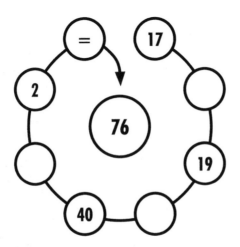

Alan Turing's father worked for the Indian Civil Service. His parents had to travel between England and India often and they left Turing and his elder brother in the care of a retired Army couple.

TILE TWISTER

Place the eight tiles into the puzzle grid so that
all adjacent numbers on each tile match up.

Tiles may be rotated through 360 degrees,
but none may be flipped over.

1	3
4	4

1	1
4	3

4	3
3	2

4	3
2	1

1	2
1	1

4	4
3	2

1	1
3	4

1	4
2	3

				2	1
				4	2

SUDOKU

Place numbers from 1 to 9 in each empty cell so that each
row, each column, and each 3x3 block contains all the
numbers from 1 to 9 to solve this tricky sudoku puzzle.

3					8		7	2
	7	9		6				
					1	9	8	
5	6					4		
		7					9	5
	3	4	5					
				4		3	6	
6	2		9					7

FUTOSHIKI

Fill the grid so that every horizontal row and
vertical column contains the numbers 1-5.

The 'greater than' or 'less than' signs indicate where a number
is larger or smaller than that in the adjacent square.

2			5	
	1			
	2			
3		5		

ONE TO NINE

Using the numbers below, complete these six equations (three reading across, and three reading downwards). Every number is used once.

1   2   3   4   5   6   7   8   9

	+		+		=	**21**
+		−		x		
	+		x		=	**20**
x		x		−		
	+		+		=	**15**
=		=		=		
72		**6**		**33**		

KAKURO

Fill the grid so that each block adds up to the
total in the box above or to the left of it.

You can only use the digits 1–9 and you must not use the same
digit twice in a block. The same digit may occur more than
once in a row or column, but it must be in a separate block.

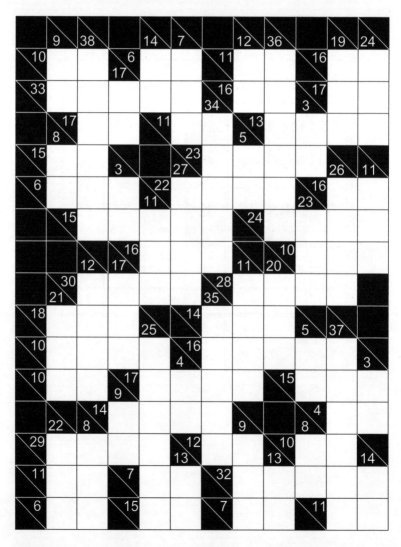

WHAT'S MISSING?

In the grid below, what number should replace the question mark?

4	13	22	31	24	17	10
17	26	35	44	37	30	23
12	21	30	39	32	25	18
20	29	38	47	40	33	26
9	18	27	36	?	22	15
32	41	50	59	52	45	38
7	16	25	34	27	20	13

DOMINO PLACEMENT

A standard set of 28 dominoes has been laid out as shown. Can you draw in the edges of them all? The check-box is provided as an aid, and the domino already placed will help.

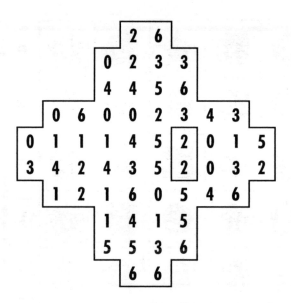

0-0	0-1	0-2	0-3	0-4	0-5	0-6	1-1	1-2	1-3	1-4	1-5	1-6	2-2
													✓

2-3	2-4	2-5	2-6	3-3	3-4	3-5	3-6	4-4	4-5	4-6	5-5	5-6	6-6

SYMBOL SUMS

Each symbol stands for a different number. In order to reach the correct total at the end of each row and column, what is the value of the circle, cross, pentagon, square, and star?

⬟	◼	★	◼	⬟	= 26
◼	◼	✚	★	✚	= 22
⬟	●	◼	★	⬟	= 27
⬟	●	●	◼	⬟	= 26
★	◼	●	⬟	◼	= 20
= 33	= 12	= 19	= 21	= 36	

TOTAL CONCENTRATION

The blank squares below should be filled with whole numbers between 1 and 30 inclusive, any of which may occur more than once, or not at all.

The numbers in every horizontal row add up to the totals on the right, as do the two long diagonal lines; whilst those in every vertical column add up to the totals along the bottom.

							103

8		27	19	4	5		95
11		12	26	30		1	118
29	15	13		16		4	119
18	23			12		8	114
25	2			3	19	20	117
	14	7	24		6	25	116
	9	13	20	26	1		88

123	102	115	137	101	104	85	77

ROUND UP

Every circle contains a number which is the sum of
the numbers in the two circles below it.

Just work out the missing numbers!

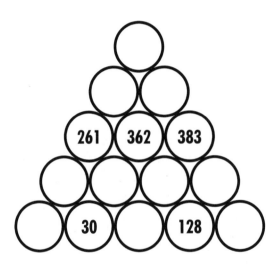

'*May not machines carry
out something which ought
to be described as thinking
but which is very different
from what a man does?*'

HEXAGONY

Can you place the hexagons into the grid, so that where any hexagon touches another along a straight line, the number in both triangles is the same?

No rotation of any hexagon is allowed!

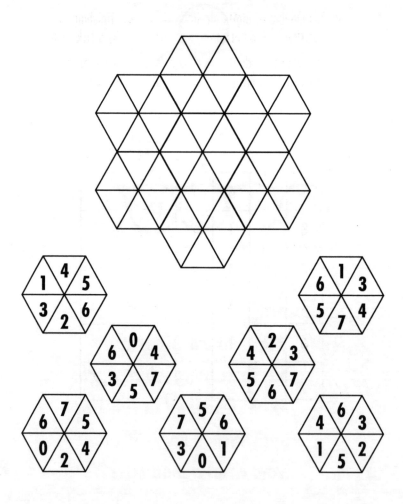

THE BOTTOM LINE

Can you fill each square in the bottom line with the correct digit?

Every square in the solution contains only one digit from each of the lines above, although two or more squares in the solution may contain the same digit.

At the end of every row is a score, which shows:

a the number of digits placed in the correct finishing position on the bottom line, as indicated by a tick; and

b the number of digits which appear on the bottom line, but in a different position, as indicated by a cross.

				SCORE
5	4	4	7	✓
8	2	1	2	✗ ✗
8	8	1	7	✗
4	5	3	1	✓
6	7	1	8	✓
				✓✓✓✓

26

COIN COLLECTING

In this puzzle, an amateur coin collector has been out with his metal detector, searching for booty. He didn't have time to dig up all the coins he found, so has made a grid map, showing their locations, in the hope that if he loses the map, at least no-one else will understand it...

Those squares containing numbers are empty, but where a number appears in a square, it indicates how many coins are located in the squares (up to a maximum of eight) surrounding the numbered one, touching it at any corner or side. There is only one coin in any individual square.

Place a circle into every square containing a coin.

1	3			3	3			0
	4			2			3	
	3						3	
4		0			1			
						2		
3		1	1			1		0
	4		2		2	1		0
			4				1	
	3					1		1
			2		2			

LATIN SQUARE

The grid should be filled with numbers from 1 to 6, so that each number appears just once in every row and column. The clues refer to the digit totals in the squares, eg A 1 2 3 = 6 means that the numbers in squares A1, A2, and A3 add up to 6.

1 C 4 5 = 8

2 D 5 6 = 7

3 E 1 2 = 7

4 F 1 2 = 8

5 A B C 1 = 10

6 A B C 2 = 6

7 C D 3 = 8

8 A B 4 = 6

9 E F 5 = 6

10 B C 6 = 8

11 A 5 6 = 9

	A	B	C	D	E	F
1						
2						
3						
4						
5						
6						

ZIGZAG

The object of this puzzle is to trace a single path
from the top left corner to the bottom right corner
of the grid, moving through all of the cells in either
a horizontal, vertical, or diagonal direction.

Every cell must be entered once only and your path should take
you through the numbers in the sequence 1-2-3-4-1-2-3-4, etc.

Can you find the way?

1	3	4	1	2	1	4	3
2	2	3	4	3	2	2	1
1	2	4	1	4	3	4	2
4	3	1	1	2	3	1	3
1	2	3	2	1	4	3	4
3	4	4	1	3	2	1	4
2	3	1	4	2	2	2	3
4	1	2	3	4	1	3	4

COMBIKU

Each horizontal row and vertical column should contain different shapes and different numbers.

Every square will contain one number and one shape, and no combination may be repeated anywhere else in the puzzle.

◇ ○ ☆ ⬡ ▢

1   2   3   4   5

①				4
		3		☆
3		▢		
☆			⬡	5
4		☆	▢	○

IT DOESN'T ADD UP

In the square below, change the positions of six numbers, one per horizontal row, vertical column, and long diagonal line of six smaller squares, in such a way that the numbers in each row, column, and long diagonal line total exactly 262.

Any number may appear more than once in a row, column, or line.

56	8	44	72	84	24
61	43	36	64	47	50
71	55	43	21	27	52
12	67	47	65	18	44
17	27	54	61	11	48
52	18	64	18	66	25

MIND OVER MATTER

Given that the letters are valued 1–26 according
to their places in the alphabet, can you crack the
mystery code to reveal the missing letter?

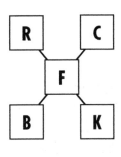

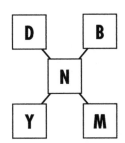

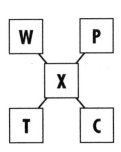

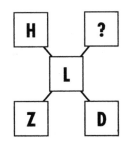

32

TILE TWISTER

Place the eight tiles into the puzzle grid so that all adjacent numbers on each tile match up.

Tiles may be rotated through 360 degrees, but none may be flipped over.

4	2
3	3

1	2
4	3

1	3
2	3

2	3
3	2

1	4
3	3

3	2
4	2

4	2
3	4

4	1
2	3

		2	1		
		4	2		

SUM CIRCLE

Fill the three empty circles with the symbols +, − and x in some order, to make a sum which totals the number in the middle. Each symbol must be used once and calculations are made in the direction of travel (clockwise).

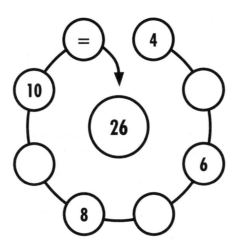

A gifted & distinguished boy, whose future career we shall watch with much interest. I have found him pleasant & friendly & I believe that he has justified his appointment as a School Prefect.

From Turing's school report, Summer Term 1931

SUDOKU

Place numbers from 1 to 9 in each empty cell so that each row, each column, and each 3x3 block contains all the numbers from 1 to 9 to solve this tricky sudoku puzzle.

				7			9	
						3	1	
					4		8	7
9		6	1			4		
3			8		2			5
		8			3	7		9
8	6		2					
	3	5						
	1			6				

FUTOSHIKI

Fill the grid so that every horizontal row and
vertical column contains the numbers 1-5.

The 'greater than' or 'less than' signs indicate where a number
is larger or smaller than that in the adjacent square.

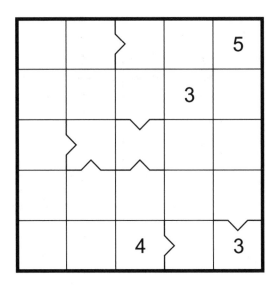

ONE TO NINE

Using the numbers below, complete these six equations (three reading across, and three reading downwards). Every number is used once.

1   2   3   4   5   6   7   8   9

	x		+		=	**29**
x	■	+	■	x		
	x		−		=	**19**
+	■	−	■	x		
	x		+		=	**55**
=		=		=		
21		**5**		**14**		

KAKURO

Fill the grid so that each block adds up to the
total in the box above or to the left of it.

You can only use the digits 1-9 and you must not use the same
digit twice in a block. The same digit may occur more than
once in a row or column, but it must be in a separate block.

38

WHAT'S MISSING?

In the grid below, what number should replace the question mark?

173	107	66	41	25	16	9
101	61	40	21	19	2	17
271	168	103	65	38	27	11
153	93	60	33	27	?	21
219	135	84	51	33	18	15
70	43	27	16	11	5	6
159	99	60	39	21	18	3

DOMINO PLACEMENT

A standard set of 28 dominoes has been laid out as shown. Can you draw in the edges of them all? The check-box is provided as an aid, and the domino already placed will help.

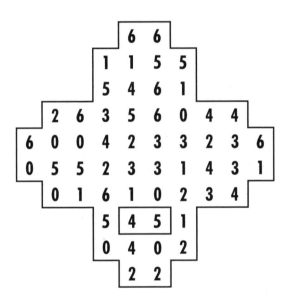

0-0	0-1	0-2	0-3	0-4	0-5	0-6	1-1	1-2	1-3	1-4	1-5	1-6	2-2

2-3	2-4	2-5	2-6	3-3	3-4	3-5	3-6	4-4	4-5	4-6	5-5	5-6	6-6
									✓				

45

ISOLATE

Draw walls to partition the grid into areas (some walls are already drawn in for you). Each area must contain two circles, area sizes must match those numbers shown next to the grid and each '+' must be linked to at least two walls.

2,   3,   3,   3,   7,   7

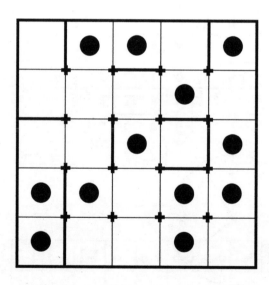

CLOCKWORK

Draw in the missing hands on the final clock.

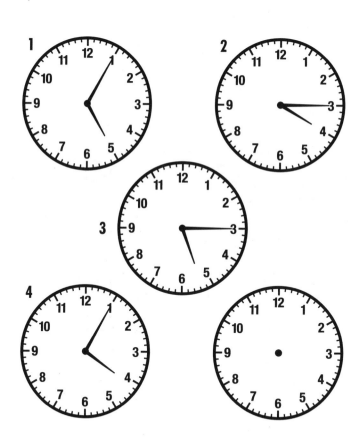

SYMBOL SUMS

Each symbol stands for a different number. In order to reach the correct total at the end of each row and column, what is the value of the circle, cross, pentagon, square, and star?

★	✚	✚	⬠	★	= 25
●	●	★	■	●	= 15
★	✚	■	■	✚	= 19
★	⬠	●	✚	●	= 21
●	★	★	⬠	⬠	= 30
= 19	= 22	= 19	= 29	= 21	

48

ROUND UP

Every circle contains a number which is the sum of the numbers in the two circles below it.

Just work out the missing numbers!

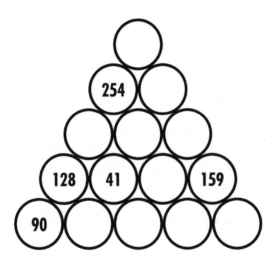

In 1936, Turing invented the idea of a machine that was able to compute anything that could be computed. This was known as the Universal Turing Machine and led to the modern computer.

HEXAGONY

Can you place the hexagons into the grid, so that where any hexagon touches another along a straight line, the number in both triangles is the same?

No rotation of any hexagon is allowed!

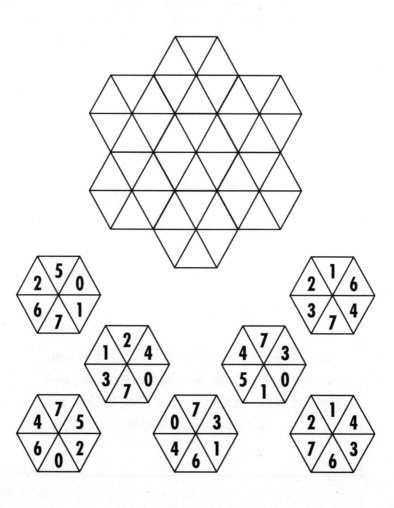

COIN COLLECTING

In this puzzle, an amateur coin collector has been out with his metal detector, searching for booty. He didn't have time to dig up all the coins he found, so has made a grid map, showing their locations, in the hope that if he loses the map, at least no-one else will understand it...

Those squares containing numbers are empty, but where a number appears in a square, it indicates how many coins are located in the squares (up to a maximum of eight) surrounding the numbered one, touching it at any corner or side. There is only one coin in any individual square.

Place a circle into every square containing a coin.

1		0	2				1	
				3		3	3	
	2					1		
0		2		3				
	1						0	
			3	2		0		
		3	2					2
2						3		5
					2		4	
	3		2					3

LATIN SQUARE

The grid should be filled with numbers from 1 to 6, so that each number appears just once in every row and column. The clues refer to the digit totals in the squares, eg A 1 2 3 = 6 means that the numbers in squares A1, A2, and A3 add up to 6.

1 B C D 2 = 6		**6** C 3 4 = 8
2 B 3 4 5 = 15		**7** D 3 4 = 5
3 E F 3 = 7		**8** A B 6 = 6
4 E 5 6 = 6		**9** E F 4 = 3
5 C D 5 = 11		**10** F 5 6 = 5

	A	B	C	D	E	F
1						
2						
3						
4						
5						
6						

ZIGZAG

The object of this puzzle is to trace a single path
from the top left corner to the bottom right corner
of the grid, moving through all of the cells in either
a horizontal, vertical, or diagonal direction.

Every cell must be entered once only and your path should take
you through the numbers in the sequence 1-2-3-4-1-2-3-4, etc.

Can you find the way?

1	2	1	2	1	2	1	4
3	4	3	3	4	3	3	2
2	4	3	4	1	2	1	3
1	2	2	4	4	3	4	2
1	4	1	3	2	1	3	1
2	3	4	1	3	2	4	2
1	2	4	4	1	1	3	3
4	3	1	3	2	2	4	4

COMBIKU

Each horizontal row and vertical column should contain different shapes and different numbers.

Every square will contain one number and one shape, and no combination may be repeated anywhere else in the puzzle.

◇ ○ ☆ ⬡ □

1   2   3   4   5

IT DOESN'T ADD UP

In the square below, change the positions of six numbers, one per horizontal row, vertical column, and long diagonal line of six smaller squares, in such a way that the numbers in each row, column, and long diagonal line total exactly 221.

Any number may appear more than once in a row, column, or line.

21	12	38	57	88	21
42	36	36	26	41	45
53	46	22	14	26	46
14	51	41	58	25	31
36	54	43	49	13	37
66	5	27	22	27	57

TILE TWISTER

Place the eight tiles into the puzzle grid so that
all adjacent numbers on each tile match up.

Tiles may be rotated through 360 degrees,
but none may be flipped over.

1	2
3	1

2	2
4	3

2	4
3	2

2	4
3	1

3	4
4	2

2	2
1	4

2	1
4	1

3	3
2	3

	4	2	
	2	2	

SUDOKU

Place numbers from 1 to 9 in each empty cell so that each
row, each column, and each 3x3 block contains all the
numbers from 1 to 9 to solve this tricky sudoku puzzle.

6		3		9				
1		4						8
					1		7	
	1				5			
		9		6		2		
			7				4	
	5		8					
2						6		1
				2		9		3

FUTOSHIKI

Fill the grid so that every horizontal row and
vertical column contains the numbers 1–5.

The 'greater than' or 'less than' signs indicate where a number
is larger or smaller than that in the adjacent square.

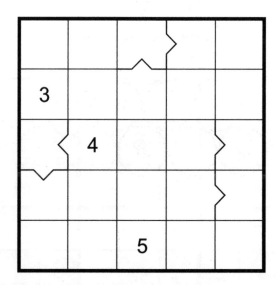

ROUND UP

Every circle contains a number which is the sum of
the numbers in the two circles below it.

Just work out the missing numbers!

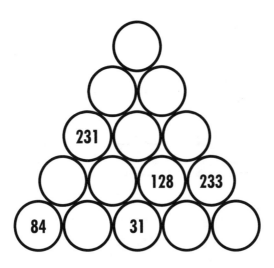

'The view that machines cannot
give rise to surprises is due,
I believe, to a fallacy to which
philosophers and mathematicians
are particularly subject.'

ONE TO NINE

Using the numbers below, complete these six equations (three reading across, and three reading downwards). Every number is used once.

1 2 3 4 5 6 7 8 9

	x		−		=	4
x	■	x	■	x		
	x		−		=	11
+	■	+	■	x		
	x		+		=	79
=		=		=		
11		39		56		

KAKURO

Fill the grid so that each block adds up to the
total in the box above or to the left of it.

You can only use the digits 1–9 and you must not use the same
digit twice in a block. The same digit may occur more than
once in a row or column, but it must be in a separate block.

56

WHAT'S MISSING?

In the grid below, what number should replace the question mark?

6	11	2	4	9	5	7
24	44	8	16	36	20	28
20	40	4	12	32	16	24
100	200	20	60	?	80	120
95	195	15	55	155	75	115
570	1170	90	330	930	450	690
564	1164	84	324	924	444	684

DOMINO PLACEMENT

A standard set of 28 dominoes has been laid out as shown. Can you draw in the edges of them all? The check-box is provided as an aid, and the domino already placed will help.

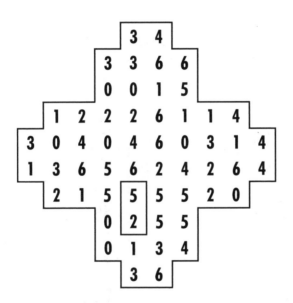

0-0	0-1	0-2	0-3	0-4	0-5	0-6	1-1	1-2	1-3	1-4	1-5	1-6	2-2

2-3	2-4	2-5	2-6	3-3	3-4	3-5	3-6	4-4	4-5	4-6	5-5	5-6	6-6
		✓											

ISOLATE

Draw walls to partition the grid into areas (some walls are already drawn in for you). Each area must contain two circles, area sizes must match those numbers shown next to the grid and each '+' must be linked to at least two walls.

2,   3,   6,   7,   7

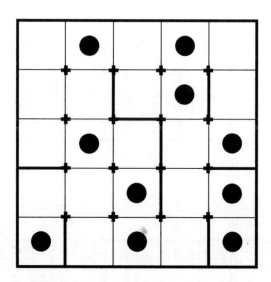

TOTAL CONCENTRATION

The blank squares below should be filled with whole numbers between 1 and 30 inclusive, any of which may occur more than once, or not at all.

The numbers in every horizontal row add up to the totals on the right, as do the two long diagonal lines; whilst those in every vertical column add up to the totals along the bottom.

							127
24	3		13	15		4	96
6	12	11		24	27		92
25		10	13	21		14	121
27	1	30	22		16		142
	4	16	23		7	19	88
	24	14	9		12	23	96
13	2			22	1	8	89
108	65	113	117	119	102	100	102

65

CLOCKWORK

Draw in the missing hands on the final clock.

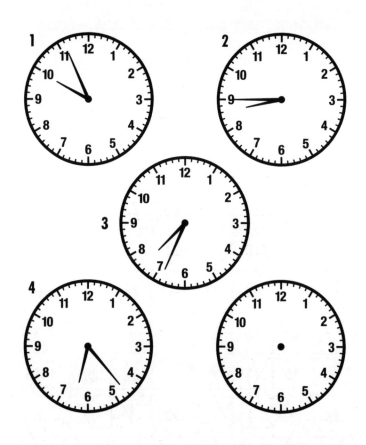

66

HEXAGONY

Can you place the hexagons into the grid, so that where any hexagon touches another along a straight line, the number in both triangles is the same?

No rotation of any hexagon is allowed!

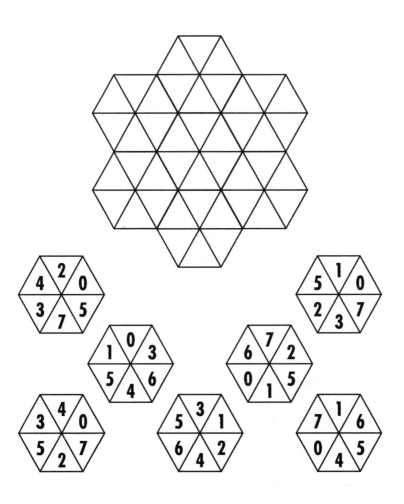

THE BOTTOM LINE

Can you fill each square in the bottom line with the correct digit?

Every square in the solution contains only one digit from each of the lines above, although two or more squares in the solution may contain the same digit.

At the end of every row is a score, which shows:

a   the number of digits placed in the correct finishing position on the bottom line, as indicated by a tick; and

b   the number of digits which appear on the bottom line, but in a different position, as indicated by a cross.

				SCORE
7	3	7	8	✓ ✗
5	8	6	3	✗
4	5	7	1	✗
1	1	2	7	✗
2	4	5	8	✓
				✓✓✓✓

SUM CIRCLE

Fill the three empty circles with the symbols +, − and x in some order, to make a sum which totals the number in the middle. Each symbol must be used once and calculations are made in the direction of travel (clockwise).

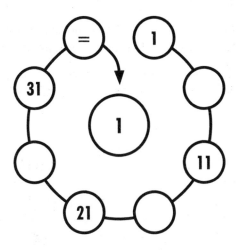

By decrypting the Nazi encryption machine "Enigma", so giving the Allies knowledge of the German army's movements, Turing affected the outcome of World War 2.

COIN COLLECTING

In this puzzle, an amateur coin collector has been out with his metal detector, searching for booty. He didn't have time to dig up all the coins he found, so has made a grid map, showing their locations, in the hope that if he loses the map, at least no-one else will understand it...

Those squares containing numbers are empty, but where a number appears in a square, it indicates how many coins are located in the squares (up to a maximum of eight) surrounding the numbered one, touching it at any corner or side. There is only one coin in any individual square.

Place a circle into every square containing a coin.

0			3					
		2			2	3	2		1
1					3		1		1
	1		1						
	2			5					
		1				2		5	
			0						
1		1		1	0		1		
		1		3				1	
	0					2			

LATIN SQUARE

The grid should be filled with numbers from 1 to 6, so that each number appears just once in every row and column. The clues refer to the digit totals in the squares, eg A 1 2 3 = 6 means that the numbers in squares A1, A2, and A3 add up to 6.

1 E F 1 = 4

6 A B 3 = 5

2 E 2 3 = 3

7 D E 4 = 7

3 B C D 2 = 9

8 E F 6 = 10

4 B 4 5 = 5

9 D 5 6 = 4

5 C 3 4 5 = 13

10 F 2 3 = 9

	A	B	C	D	E	F
1						
2						
3						
4						
5						
6						

ZIGZAG

The object of this puzzle is to trace a single path
from the top left corner to the bottom right corner
of the grid, moving through all of the cells in either
a horizontal, vertical, or diagonal direction.

Every cell must be entered once only and your path should take
you through the numbers in the sequence 1-2-3-4-1-2-3-4, etc.

Can you find the way?

1	2	1	4	3	4	3	2
1	2	3	2	1	4	2	1
4	3	4	3	2	3	1	4
2	1	4	4	1	2	4	3
3	1	2	3	3	3	2	1
2	4	4	2	1	4	1	3
1	3	1	1	4	1	4	2
4	3	2	2	3	2	3	4

COMBIKU

Each horizontal row and vertical column should
contain different shapes and different numbers.

Every square will contain one number and one shape, and no
combination may be repeated anywhere else in the puzzle.

◇	◯	☆	⬡	▢
1	2	3	4	5

1	⬡	5	◯	
◯	▢	◇		3
	5	▢	2	
			1	
3				5

IT DOESN'T ADD UP

In the square below, change the positions of six numbers, one per horizontal row, vertical column, and long diagonal line of six smaller squares, in such a way that the numbers in each row, column, and long diagonal line total exactly 121.

Any number may appear more than once in a row, column, or line.

13	13	15	23	31	29
30	20	10	10	21	19
26	38	20	4	19	20
15	19	21	24	11	19
13	22	16	34	17	36
21	12	28	14	28	15

TILE TWISTER

Place the eight tiles into the puzzle grid so that
all adjacent numbers on each tile match up.

Tiles may be rotated through 360 degrees,
but none may be flipped over.

1	2
4	3

1	2
3	4

1	3
4	3

4	1
3	4

2	1
2	4

3	1
2	3

1	1
2	4

4	4
2	3

2	3		
4	3		

SUDOKU

Place numbers from 1 to 9 in each empty cell so that each row, each column, and each 3x3 block contains all the numbers from 1 to 9 to solve this tricky sudoku puzzle.

	5	8			4			
3	1			7			8	2
						6		
	9				8	5		
			6		7			
		2	3				4	
		1						
9	4			5			2	8
			1			7	3	

FUTOSHIKI

Fill the grid so that every horizontal row and
vertical column contains the numbers 1-5.

The 'greater than' or 'less than' signs indicate where a number
is larger or smaller than that in the adjacent square.

		4	3	
	3		<	
∧	>		5	
	2			
∨				

72

MIND OVER MATTER

Given that the letters are valued 1–26 according to their places in the alphabet, can you crack the mystery code to reveal the missing letter?

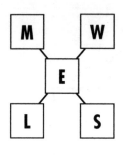

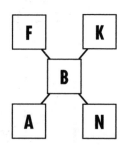

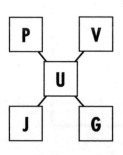

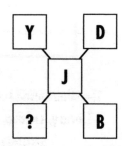

SUM CIRCLE

Fill the three empty circles with the symbols +, − and x in some order, to make a sum which totals the number in the middle. Each symbol must be used once and calculations are made in the direction of travel (clockwise).

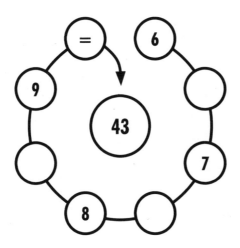

The Alan Turing Institute, headquartered in the British Library, London, was created as the UK's national institute for data science in 2015.

ONE TO NINE

Using the numbers below, complete these six equations (three reading across, and three reading downwards). Every number is used once.

1 2 3 4 5 6 7 8 9

	x		x		=	126
x		−		+		
	−		x		=	20
−		x		x		
	x		x		=	30
=		=		=		
71		20		36		

KAKURO

Fill the grid so that each block adds up to the
total in the box above or to the left of it.

You can only use the digits 1–9 and you must not use the same
digit twice in a block. The same digit may occur more than
once in a row or column, but it must be in a separate block.

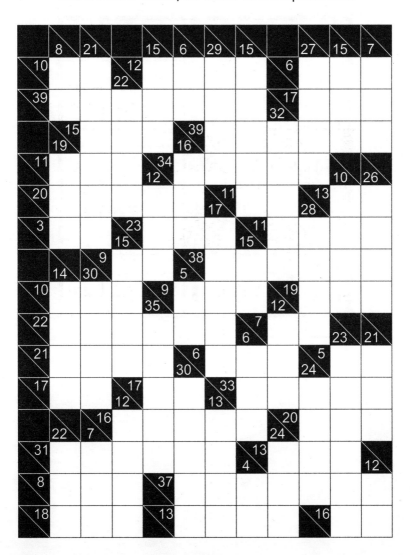

76

WHAT'S MISSING?

In the grid below, what number should replace the question mark?

3	10	17	24	31	38	45
164	171	178	185	192	199	52
157	276	283	290	297	206	59
150	269	332	?	304	213	66
143	262	325	318	311	220	73
136	255	248	241	234	227	80
129	122	115	108	101	94	87

DOMINO PLACEMENT

A standard set of 28 dominoes has been laid out as shown. Can you draw in the edges of them all? The check-box is provided as an aid, and the domino already placed will help.

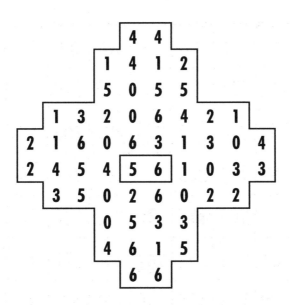

0-0	0-1	0-2	0-3	0-4	0-5	0-6	1-1	1-2	1-3	1-4	1-5	1-6	2-2

2-3	2-4	2-5	2-6	3-3	3-4	3-5	3-6	4-4	4-5	4-6	5-5	5-6	6-6
												✔	

ISOLATE

Draw walls to partition the grid into areas (some walls are already drawn in for you). Each area must contain two circles, area sizes must match those numbers shown next to the grid and each '+' must be linked to at least two walls.

2,   3,   6,   7,   7

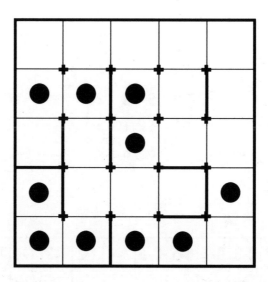

CLOCKWORK

Draw in the missing hands on the final clock.

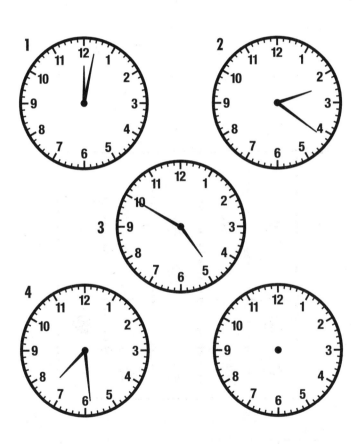

85

80

TOTAL CONCENTRATION

The blank squares below should be filled with whole numbers between 1 and 30 inclusive, any of which may occur more than once, or not at all.

The numbers in every horizontal row add up to the totals on the right, as do the two long diagonal lines; whilst those in every vertical column add up to the totals along the bottom.

							84

13		9	7	22	21		**101**
17	4		29		3	9	**108**
15		16	22	28		10	**116**
20	13	21			7	26	**119**
	11		25	16	23	5	**94**
27	6	19			10	30	**115**
	22	16	15	12	4		**98**
97	**91**	**120**	**126**	**124**	**69**	**124**	**99**

THE BOTTOM LINE

Can you fill each square in the bottom line with the correct digit?

Every square in the solution contains only one digit from each of the lines above, although two or more squares in the solution may contain the same digit.

At the end of every row is a score, which shows:

a   the number of digits placed in the correct finishing position on the bottom line, as indicated by a tick; and

b   the number of digits which appear on the bottom line, but in a different position, as indicated by a cross.

SCORE

8	5	2	2	✓ ✗ ✗
5	6	6	7	✓
6	2	4	2	✗ ✗
4	4	1	7	✓
3	4	8	2	✗ ✗
				✓ ✓ ✓ ✓

HEXAGONY

Can you place the hexagons into the grid, so that where any hexagon touches another along a straight line, the number in both triangles is the same?

No rotation of any hexagon is allowed!

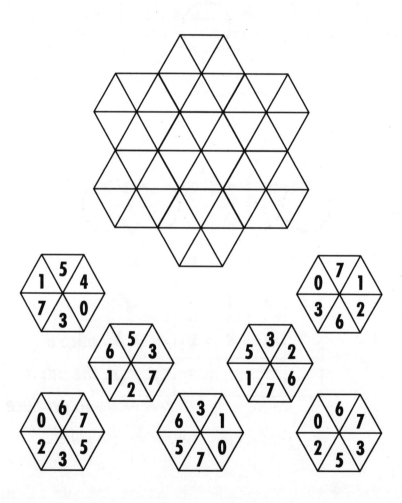

ROUND UP

Every circle contains a number which is the sum of the numbers in the two circles below it.

Just work out the missing numbers!

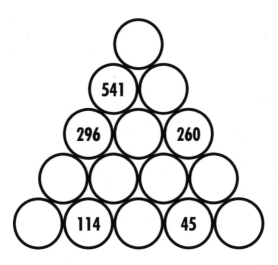

'Instead of trying to produce a programme to simulate the adult mind, why not rather try to produce one which simulates the child's?'

COIN COLLECTING

In this puzzle, an amateur coin collector has been out with his metal detector, searching for booty. He didn't have time to dig up all the coins he found, so has made a grid map, showing their locations, in the hope that if he loses the map, at least no-one else will understand it...

Those squares containing numbers are empty, but where a number appears in a square, it indicates how many coins are located in the squares (up to a maximum of eight) surrounding the numbered one, touching it at any corner or side. There is only one coin in any individual square.

Place a circle into every square containing a coin.

1		1			0			
			0					3
3			1			0		
	3					0		
	4		5		2	0		2
	3				1		3	
			4			2		
3		2	1					
	3		2		1		3	3
			0				2	

LATIN SQUARE

The grid should be filled with numbers from 1 to 6, so that each number appears just once in every row and column. The clues refer to the digit totals in the squares, eg A 1 2 3 = 6 means that the numbers in squares A1, A2, and A3 add up to 6.

1 C D E 3 = 9   7 B 1 2 = 8

2 C 4 5 = 11   8 F 4 5 = 5

3 D 4 5 6 = 13   9 A B 6 = 7

4 A B 4 = 8   10 C D 1 = 6

5 A B 5 = 6   11 E 5 6 = 6

6 A 1 2 = 7   12 C D 2 = 7

	A	B	C	D	E	F
1						
2						
3						
4						
5						
6						

ZIGZAG

The object of this puzzle is to trace a single path from the top left corner to the bottom right corner of the grid, moving through all of the cells in either a horizontal, vertical, or diagonal direction.

Every cell must be entered once only and your path should take you through the numbers in the sequence 1-2-3-4-1-2-3-4, etc.

Can you find the way?

1	2	2	3	2	3	1	2
3	1	4	3	4	1	4	3
4	1	2	2	3	1	4	1
4	1	1	4	4	2	2	3
2	3	2	3	3	3	2	4
2	1	1	4	3	1	4	1
4	3	1	4	4	2	1	2
3	2	4	1	2	3	3	4

COMBIKU

Each horizontal row and vertical column should contain different shapes and different numbers.

Every square will contain one number and one shape, and no combination may be repeated anywhere else in the puzzle.

◇ ○ ☆ ⬡ ▢

1 **2** **3** **4** **5**

		1		3
	☆		4	
		◇		
		5		
4	2		①	◇

IT DOESN'T ADD UP

In the square below, change the positions of six numbers,
one per horizontal row, vertical column, and long diagonal
line of six smaller squares, in such a way that the numbers in
each row, column, and long diagonal line total exactly 195.

Any number may appear more than once in a row, column, or line.

17	33	21	32	39	45
49	33	32	14	35	33
45	48	32	18	19	16
18	42	31	46	22	32
35	12	46	39	49	33
50	28	29	38	40	19

TILE TWISTER

Place the eight tiles into the puzzle grid so that all adjacent numbers on each tile match up.

Tiles may be rotated through 360 degrees, but none may be flipped over.

1	4
4	1

1	1
3	2

4	3
4	3

2	4
3	1

1	1
3	4

2	4
4	3

3	4
3	1

1	4
2	1

				4	1
				2	2

95

90

SUDOKU

Place numbers from 1 to 9 in each empty cell so that each row, each column, and each 3x3 block contains all the numbers from 1 to 9 to solve this tricky sudoku puzzle.

		2			9			
				3		6		
4						5		
8								2
	1			5			4	
3								9
		9						7
		6		1				
			4			8		

FUTOSHIKI

Fill the grid so that every horizontal row and
vertical column contains the numbers 1-5.

The 'greater than' or 'less than' signs indicate where a number
is larger or smaller than that in the adjacent square.

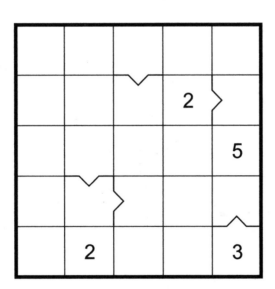

MIND OVER MATTER

Given that the letters are valued 1-26 according to their places in the alphabet, can you crack the mystery code to reveal the missing letter?

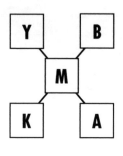

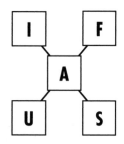

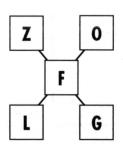

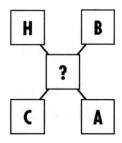

SUM CIRCLE

Fill the three empty circles with the symbols +, − and x in some order, to make a sum which totals the number in the middle. Each symbol must be used once and calculations are made in the direction of travel (clockwise).

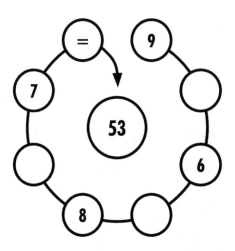

In 2018 researchers in China used theories from Turing's sole chemistry paper to develop a highly efficient water filter.

94

ONE TO NINE

Using the numbers below, complete these six equations (three reading across, and three reading downwards). Every number is used once.

1 2 3 4 5 6 7 8 9

	x		x		=	84
−	■	x	■	x		
	+		x		=	40
x	■	−	■	−		
	x		+		=	32
=		=		=		
9		25		43		

SPOT NUMBERS

The numbers at the top and on the left side show
the quantity of single-digit numbers (1-9) used in
that row and column. The numbers at the bottom
and on the right show the sum of the digits.

Any number may appear more than once in a row or column,
but no numbers are in squares that touch, even at a corner.

	2	1	3	0	3	0	3	
1								8
2								6
2	6							11
2								14
1								3
0								0
4					6			25
	15	3	19	0	12	0	18	

WHAT'S MISSING?

In the grid below, what number should replace the question mark?

26	30	23	27	20	24	17
12	19	15	22	18	25	21
9	13	6	10	3	7	?
4	11	7	14	10	17	13
11	15	8	12	5	9	2
17	24	20	27	23	30	26
18	22	15	19	12	16	9

SYMBOL SUMS

Each symbol stands for a different number. In order to reach the correct total at the end of each row and column, what is the value of the circle, cross, pentagon, square, and star?

✚	★	⬠	●	✚	= 32
⬠	✚	◼	✚	⬠	= 25
✚	◼	★	✚	✚	= 38
★	◼	●	●	★	= 27
●	◼	✚	★	◼	= 27
=	=	=	=	=	
32	26	26	34	31	

103

DOMINO PLACEMENT

A standard set of 28 dominoes has been laid out as shown. Can you draw in the edges of them all? The check-box is provided as an aid, and the domino already placed will help.

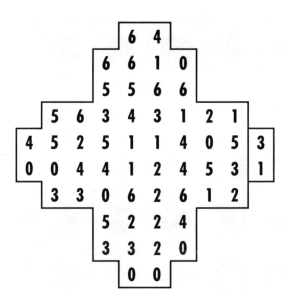

0-0	0-1	0-2	0-3	0-4	0-5	0-6	1-1	1-2	1-3	1-4	1-5	1-6	2-2
									✔				

2-3	2-4	2-5	2-6	3-3	3-4	3-5	3-6	4-4	4-5	4-6	5-5	5-6	6-6

ISOLATE

Draw walls to partition the grid into areas (some walls are already drawn in for you). Each area must contain two circles, area sizes must match those numbers shown next to the grid and each '+' must be linked to at least two walls.

3, 3, 5, 7, 7

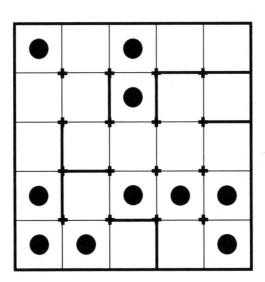

CLOCKWORK

Draw in the missing hands on the final clock.

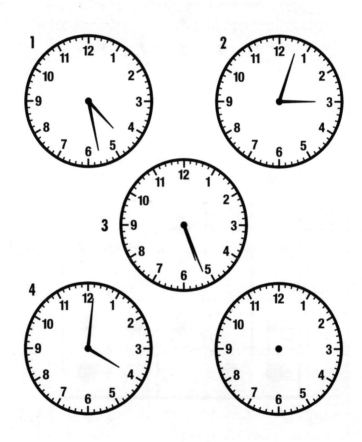

TOTAL CONCENTRATION

The blank squares below should be filled with whole numbers between 1 and 30 inclusive, any of which may occur more than once, or not at all.

The numbers in every horizontal row add up to the totals on the right, as do the two long diagonal lines; whilst those in every vertical column add up to the totals along the bottom.

							83

		19	6	17		29	**106**
17	14		20		1	11	**91**
22	30	2		13	9		**104**
16		24		13	16	4	**80**
	25	17	10	21		7	**109**
14	8		1	15	16		**87**
	17	6	23	29		2	**120**
89	**122**	**113**	**80**	**115**	**106**	**72**	**61**

HEXAGONY

Can you place the hexagons into the grid, so that where any hexagon touches another along a straight line, the number in both triangles is the same?

No rotation of any hexagon is allowed!

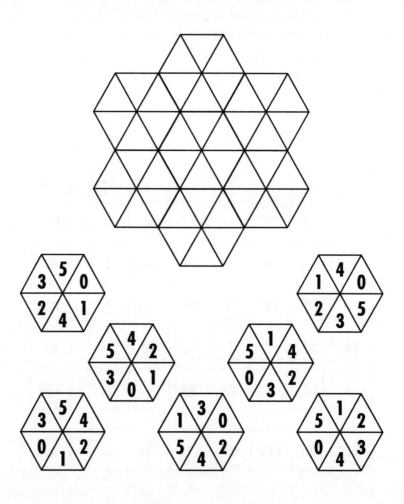

THE BOTTOM LINE

Can you fill each square in the bottom line with the correct digit?

Every square in the solution contains only one digit from each of the lines above, although two or more squares in the solution may contain the same digit.

At the end of every row is a score, which shows:

a   the number of digits placed in the correct finishing position on the bottom line, as indicated by a tick; and

b   the number of digits which appear on the bottom line, but in a different position, as indicated by a cross.

SCORE

2	6	7	7	✓
5	4	1	1	✗
6	6	9	3	✗
7	3	9	5	✓ ✗
7	3	5	4	✓
				✓✓✓✓

104

COIN COLLECTING

In this puzzle, an amateur coin collector has been out with his metal detector, searching for booty. He didn't have time to dig up all the coins he found, so has made a grid map, showing their locations, in the hope that if he loses the map, at least no-one else will understand it...

Those squares containing numbers are empty, but where a number appears in a square, it indicates how many coins are located in the squares (up to a maximum of eight) surrounding the numbered one, touching it at any corner or side. There is only one coin in any individual square.

Place a circle into every square containing a coin.

				1				2	
	1	2			1	2			2
			3		2	1	3		
1	1								
1				3			3	3	3
		1		3					2
		2		3		3		3	
	2			3	1				
0			2	1				2	
							3		

LATIN SQUARE

The grid should be filled with numbers from 1 to 6, so that each number appears just once in every row and column. The clues refer to the digit totals in the squares, eg A 1 2 3 = 6 means that the numbers in squares A1, A2, and A3 add up to 6.

1 C 4 5 = 7

2 D E F 4 = 8

3 D E F 5 = 11

4 D 1 2 = 9

5 E 1 2 = 8

6 F 1 2 = 11

7 B C 1 = 5

8 C D 6 = 8

9 A B 2 = 5

10 A 3 4 = 11

11 B 5 6 = 9

	A	B	C	D	E	F
1						
2						
3						
4						
5						
6						

ZIGZAG

The object of this puzzle is to trace a single path
from the top left corner to the bottom right corner
of the grid, moving through all of the cells in either
a horizontal, vertical, or diagonal direction.

Every cell must be entered once only and your path should take
you through the numbers in the sequence 1-2-3-4-1-2-3-4, etc.

Can you find the way?

1	3	2	1	2	4	1	3
2	1	4	3	3	2	4	2
1	3	4	1	4	1	3	1
4	2	2	3	4	2	3	4
3	4	2	3	2	1	4	2
1	2	1	1	4	3	3	1
1	3	4	2	4	2	3	4
4	2	3	3	1	2	1	4

COMBIKU

Each horizontal row and vertical column should contain different shapes and different numbers.

Every square will contain one number and one shape, and no combination may be repeated anywhere else in the puzzle.

◇   ○   ☆   ⬡   ▢

1   2   3   4   5

		☆		
		▢		④
		5	**3**	**2**
	⬡3			
4	**2**	⬡		

IT DOESN'T ADD UP

In the square below, change the positions of six numbers, one per horizontal row, vertical column, and long diagonal line of six smaller squares, in such a way that the numbers in each row, column, and long diagonal line total exactly 145.

Any number may appear more than once in a row, column, or line.

29	27	34	21	19	25
24	24	20	18	25	33
19	29	24	26	20	19
22	31	25	14	17	25
22	8	26	33	36	17
28	39	26	25	25	15

TILE TWISTER

Place the eight tiles into the puzzle grid so that all adjacent numbers on each tile match up.

Tiles may be rotated through 360 degrees, but none may be flipped over.

1	3
1	2

2	3
4	3

2	4
4	4

1	3
4	2

2	2
4	3

4	3
2	1

1	2
3	1

4	2
4	2

SUDOKU

Place numbers from 1 to 9 in each empty cell so that each row, each column, and each 3x3 block contains all the numbers from 1 to 9 to solve this tricky sudoku puzzle.

	7			2				
		1						9
		6				8		5
					5			
	4						7	
				1	9			
		9						
			7				6	
2		5						

FUTOSHIKI

Fill the grid so that every horizontal row and
vertical column contains the numbers 1-5.

The 'greater than' or 'less than' signs indicate where a number
is larger or smaller than that in the adjacent square.

	>			∧
1				
5	4	<	2	
4	1	5		

ONE TO NINE

Using the numbers below, complete these six
equations (three reading across, and three reading
downwards). Every number is used once.

1 2 3 4 5 6 7 8 9

	+		–		=	8
–	■	+	■	x		
	–		x		=	8
+	■	x	■	+		
	x		x		=	10
=		=		=		
4		12		61		

118

SPOT NUMBERS

The numbers at the top and on the left side show the quantity of single-digit numbers (1-9) used in that row and column. The numbers at the bottom and on the right show the sum of the digits.

Any number may appear more than once in a row or column, but no numbers are in squares that touch, even at a corner.

	2	1	2	1	3	0	3	
3	9							21
1								4
1								2
1								4
2					6			9
1								4
3								17
	10	4	10	2	21	0	14	

114

WHAT'S MISSING?

In the grid below, what number should replace the question mark?

72	61	49	38	26	15	3
91	79	66	54	41	29	16
106	93	79	66	52	39	25
94	80	65	51	36	22	7
111	96	80	65	49	34	18
121	105	88	72	55	39	22
?	96	78	61	43	26	8

SYMBOL SUMS

Each symbol stands for a different number. In order to reach the correct total at the end of each row and column, what is the value of the circle, cross, pentagon, square, and star?

✚	★	⬟	■	⬟	= 25
★	★	✚	■	■	= 31
●	✚	★	⬟	■	= 27
✚	■	★	★	●	= 32
✚	★	✚	■	■	= 30
=	=	=	=	=	
34	**35**	**33**	**23**	**20**	

121

DOMINO PLACEMENT

A standard set of 28 dominoes has been laid out as shown. Can you draw in the edges of them all? The check-box is provided as an aid, and the domino already placed will help.

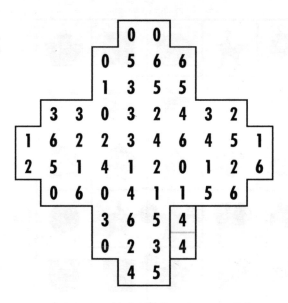

0-0	0-1	0-2	0-3	0-4	0-5	0-6	1-1	1-2	1-3	1-4	1-5	1-6	2-2

2-3	2-4	2-5	2-6	3-3	3-4	3-5	3-6	4-4	4-5	4-6	5-5	5-6	6-6
								✓					

ISOLATE

Draw walls to partition the grid into areas (some walls are already drawn in for you). Each area must contain two circles, area sizes must match those numbers shown next to the grid and each '+' must be linked to at least two walls.

2, 3, 4, 4, 5, 7

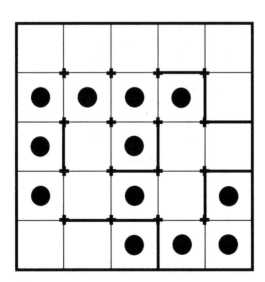

CLOCKWORK

Draw in the missing hands on the final clock.

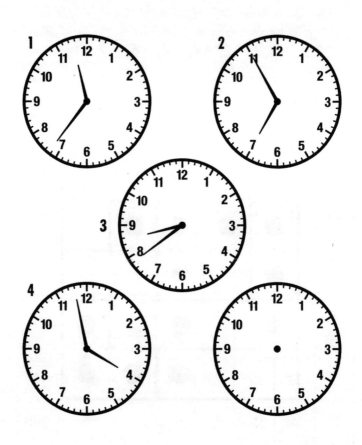

TOTAL CONCENTRATION

The blank squares below should be filled with whole numbers between 1 and 30 inclusive, any of which may occur more than once, or not at all.

The numbers in every horizontal row add up to the totals on the right, as do the two long diagonal lines; whilst those in every vertical column add up to the totals along the bottom.

							98
20			17	8		7	87
24	16	9	4		26		85
3	18			11	12	13	89
	6	13	19		20	2	108
10				29	24	9	116
	12	6	27	19	10	16	104
15	20	21	17	4			100
113	101	93	108	93	117	64	136

HEXAGONY

Can you place the hexagons into the grid, so that where any hexagon touches another along a straight line, the number in both triangles is the same?

No rotation of any hexagon is allowed!

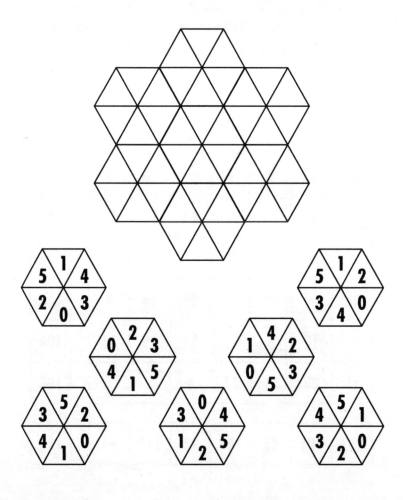

THE BOTTOM LINE

Can you fill each square in the bottom line with the correct digit?

Every square in the solution contains only one digit from each of the lines above, although two or more squares in the solution may contain the same digit.

At the end of every row is a score, which shows:

a   the number of digits placed in the correct finishing position on the bottom line, as indicated by a tick; and

b   the number of digits which appear on the bottom line, but in a different position, as indicated by a cross.

				SCORE
1	1	4	2	✓ ✗ ✗
1	5	2	4	✓ ✗
2	2	7	6	✓
0	6	4	3	✗
4	7	1	7	✓
				✓✓✓✓

COIN COLLECTING

In this puzzle, an amateur coin collector has been out with his metal detector, searching for booty. He didn't have time to dig up all the coins he found, so has made a grid map, showing their locations, in the hope that if he loses the map, at least no-one else will understand it...

Those squares containing numbers are empty, but where a number appears in a square, it indicates how many coins are located in the squares (up to a maximum of eight) surrounding the numbered one, touching it at any corner or side. There is only one coin in any individual square.

Place a circle into every square containing a coin.

2		1				3			
		2	2						3
3				3	3	2			1
2		2		1					2
			2		2			1	
			2			1		1	
	2		1			2		1	
	2		2			2			
	2			4			1		
	1				1	2		1	

LATIN SQUARE

The grid should be filled with numbers from 1 to 6, so that each number appears just once in every row and column. The clues refer to the digit totals in the squares, eg A 1 2 3 = 6 means that the numbers in squares A1, A2, and A3 add up to 6.

1	D 1 2 = 7	**7**	C 1 2 = 5
2	A B 1 = 6	**8**	F 1 2 3 = 11
3	A B 2 = 11	**9**	D E 6 = 9
4	B 4 5 6 = 11	**10**	A 3 4 = 3
5	C D 5 = 5	**11**	E 1 2 = 5
6	C D 4 = 8	**12**	B C 3 = 9

	A	B	C	D	E	F
1						
2						
3						
4						
5						
6						

ZIGZAG

The object of this puzzle is to trace a single path
from the top left corner to the bottom right corner
of the grid, moving through all of the cells in either
a horizontal, vertical, or diagonal direction.

Every cell must be entered once only and your path should take
you through the numbers in the sequence 1-2-3-4-1-2-3-4, etc.

Can you find the way?

1	3	4	1	2	1	4	3
2	4	2	3	3	4	2	2
3	4	1	3	2	3	1	1
2	1	2	4	1	3	4	2
2	1	3	1	4	4	1	3
3	4	2	3	2	1	4	2
4	3	2	1	4	3	1	3
1	2	4	1	4	3	2	4

COMBIKU

Each horizontal row and vertical column should contain different shapes and different numbers.

Every square will contain one number and one shape, and no combination may be repeated anywhere else in the puzzle.

◇ ○ ☆ ⬡ ▢

1   2   3   4   5

4				
	☆3		◇4	▢
◇5		1		
▢		☆4		2
		3		⬡

IT DOESN'T ADD UP

In the square below, change the positions of six numbers, one per horizontal row, vertical column, and long diagonal line of six smaller squares, in such a way that the numbers in each row, column, and long diagonal line total exactly 196.

Any number may appear more than once in a row, column, or line.

53	15	23	28	38	30
52	32	29	23	36	36
31	66	32	34	15	32
31	52	36	40	13	34
22	32	39	46	28	34
19	13	42	35	34	21

TILE TWISTER

Place the eight tiles into the puzzle grid so that all adjacent numbers on each tile match up.

Tiles may be rotated through 360 degrees, but none may be flipped over.

2	4
2	1

1	3
1	2

4	4
2	3

1	1
4	4

1	4
4	2

3	2
2	1

1	3
4	1

2	1
4	3

SUDOKU

Place numbers from 1 to 9 in each empty cell so that each row, each column, and each 3x3 block contains all the numbers from 1 to 9 to solve this tricky sudoku puzzle.

						2	9	
			6					
			1		7			
		8						
	1		3					
				2		8		
5							3	
6		4		5				
							1	7

FUTOSHIKI

Fill the grid so that every horizontal row and
vertical column contains the numbers 1–5.

The 'greater than' or 'less than' signs indicate where a number
is larger or smaller than that in the adjacent square.

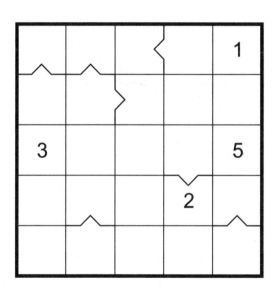

ONE TO NINE

Using the numbers below, complete these six equations (three reading across, and three reading downwards). Every number is used once.

1 2 3 4 5 6 7 8 9

	+		−		=	5
x	■	−	■	x		
	+		+		=	16
x	■	x	■	−		
	x		−		=	27
=		=		=		
135		21		4		

SPOT NUMBERS

The numbers at the top and on the left side show the quantity of single-digit numbers (1-9) used in that row and column. The numbers at the bottom and on the right show the sum of the digits.

Any number may appear more than once in a row or column, but no numbers are in squares that touch, even at a corner.

	4	0	3	1	1	1	2	
3			1					**15**
0								**0**
3								**16**
1								**1**
2								**9**
1								**2**
2								**11**
	10	**0**	**16**	**9**	**8**	**8**	**3**	

132

WHAT'S MISSING?

In the grid below, what number should replace the question mark?

26	15	22	34	19	28	11
30	21	26	40	23	34	15
24	17	20	36	17	30	9
28	23	24	42	21	36	13
22	19	18	38	15	32	7
26	25	22	44	19	38	11
20	21	16	40	13	34	?

SYMBOL SUMS

Each symbol stands for a different number. In order to reach the correct total at the end of each row and column, what is the value of the circle, cross, pentagon, square, and star?

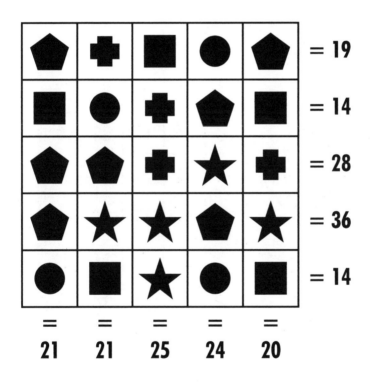

1

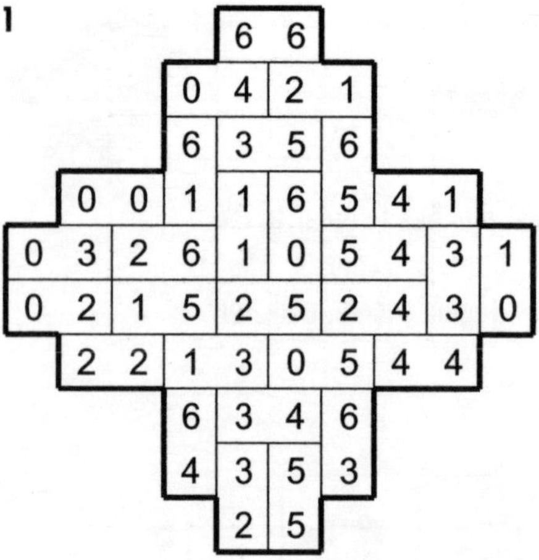

2

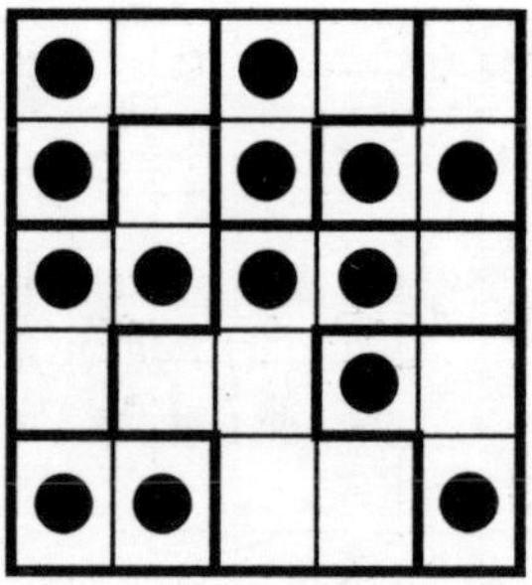

3

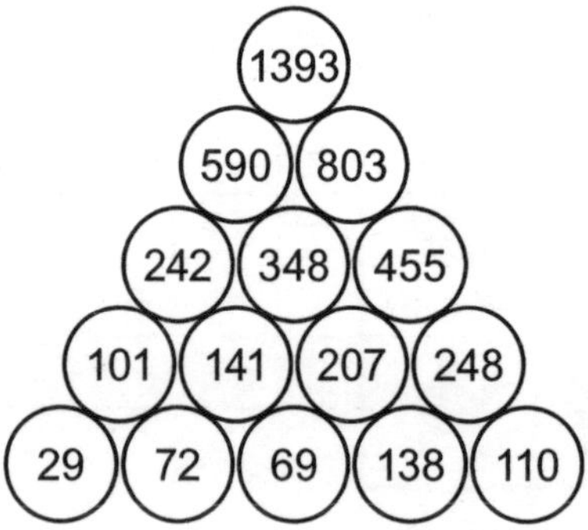

4

							70
28	16	2	18	21	24	5	114
6	14	17	20	9	1	19	86
23	12	2	11	13	15	30	106
22	8	27	18	17	12	8	112
2	4	7	29	21	25	18	106
5	10	26	3	12	14	28	98
16	22	21	1	17	18	9	104
102	86	102	100	110	109	117	106

5

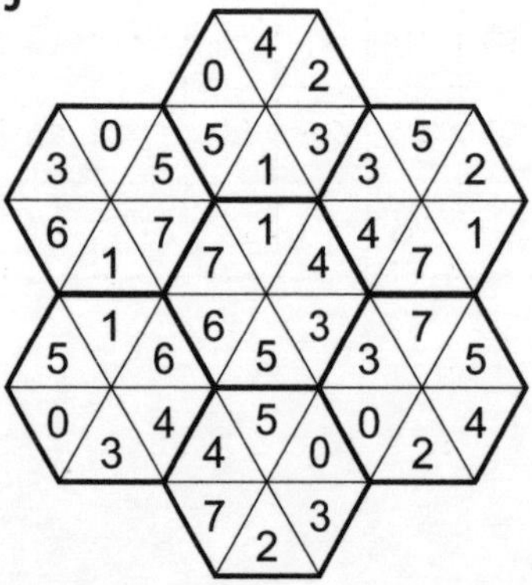

6

4874

7

		1				●		●	
1	3	●	3	●		1	2	1	
2	●	●			2				1
3	●	●	4	●	3		●	4	●
●				●	●		●	4	●
2		0		3					1
●			3	●		0			
●	●	●	●	2					
2						2	2	1	
	0		●	1	1	●	●		

8

1	2	4	3	6	5
6	3	5	2	1	4
5	1	3	4	2	6
2	4	6	1	5	3
3	6	1	5	4	2
4	5	2	6	3	1

9

1	2	2	1	4	1	3	2
4	3	8	3	2	4	4	1
1	4	1	3	3	1	3	2
2	4	4	2	2	4	3	1
3	1	2	1	4	1	2	4
4	1	3	1	2	3	4	3
3	2	4	2	1	2	3	2
2	1	3	4	3	4	1	4

10

1	5	4	2	3
5	3	2	1	4
4	2	5	3	1
2	1	3	4	5
3	4	1	5	2

11

39	13	24	63	68	39
38	41	66	15	41	45
41	74	41	23	21	46
33	49	41	59	20	44
36	43	25	58	39	45
59	26	49	28	57	27

12

The value of the letter in the central square is the square root of the sum total of the values of the letters in the outer squares. Thus the missing value is 6, so the missing letter is F.

13

17 − 19 + 40 x 2 = 76

14

1	1	1	2	2	1
4	3	3	4	4	2
4	3	3	4	4	2
1	1	1	4	4	3
1	1	1	4	4	3
1	2	2	3	3	2

15

3	1	5	4	9	8	6	7	2
8	7	9	3	6	2	1	5	4
2	4	6	7	5	1	9	8	3
5	6	2	1	7	9	4	3	8
4	9	3	8	2	5	7	1	6
1	8	7	6	3	4	2	9	5
7	3	4	5	1	6	8	2	9
9	5	8	2	4	7	3	6	1
6	2	1	9	8	3	5	4	7

16

2	3	1	5	4
5	1	4	3	2
1	2	3	4	5
3	4	5	2	1
4	5	2	1	3

17

9	+	4	+	8	=	21
+		−		x		
3	+	1	x	5	=	20
x		x		−		
6	+	2	+	7	=	15
=		=		=		
72		6		33		

19

29 – Reading along each row, add 9 to each preceding number until the central number, after which deduct 7 from each preceding number.

18

4	6		5	1		3	8		7	9
5	7	8	9	4		9	7		9	8
	8	9		2	9		2	1	3	7
6	9			8	4	9	2			
2	3	1		8	7	1	6		9	7
	5	2	1	3	4		4	9	8	3
		3	7	6			6	3	1	
	6	8	7	9		5	9	8	6	
9	3	6			5	1	8			
4	1	3	2		6	2	3	1	4	
8	2		4	1	9	3		4	9	2
		2	1	3	8			3	1	
8	5	7	9		7	5		3	7	
9	2		3	4		3	7	5	8	9
5	1		6	9		1	6		6	5

20

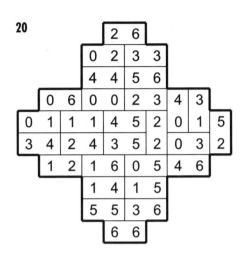

21

Circle = 3, cross = 7,
pentagon = 9, square = 2,
star = 4.

22

							103
8	22	27	19	4	5	10	95
11	17	12	26	30	21	1	118
29	15	13	14	16	28	4	119
18	23	16	13	12	24	8	114
25	2	27	21	3	19	20	117
30	14	7	24	10	6	25	116
2	9	13	20	26	1	17	88
123	102	115	137	101	104	85	77

23

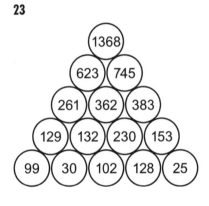

24

25

5528

26

1	3	●		●	3	3	●		0
●	4	●		2	●	●		3	
	●	3					●	3	●
4	●		0			1		●	
●	●						2		
3		1	1			1	●		0
●	4		2	●	2		1		0
●	●	●	4		●			1	
	3		●	●	●		1	●	1
			2		2				

27

1	5	4	6	3	2
2	3	1	5	4	6
5	4	6	2	1	3
4	2	3	1	6	5
6	1	5	3	2	4
3	6	2	4	5	1

28

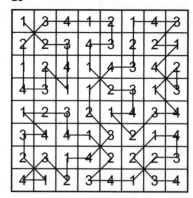

29

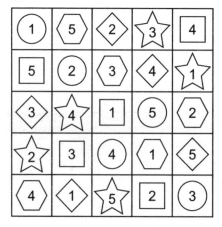

30

56	8	18	72	84	24
61	43	36	25	47	50
64	55	43	21	27	52
12	67	47	65	27	44
17	71	54	61	11	48
52	18	64	18	66	44

31

The value of the central letter is the sum total of the letters in the left squares minus the sum total of the letters in the right squares. Thus the missing value is 18, so the missing letter is R.

32

2	4	4	2	2	3
2	3	3	4	4	3
2	3	3	4	4	3
3	2	2	1	1	3
3	2	2	1	1	3
1	4	4	2	2	3

33

4 x 6 – 8 + 10 = 26

34

4	8	3	5	7	1	6	9	2
5	7	2	9	8	6	3	1	4
6	9	1	3	2	4	5	8	7
9	2	6	1	5	7	4	3	8
3	4	7	8	9	2	1	6	5
1	5	8	6	4	3	7	2	9
8	6	4	2	3	5	9	7	1
2	3	5	7	1	9	8	4	6
7	1	9	4	6	8	2	5	3

35

3	2 >	1	4	5
1	4	5	3	2
4 >	3	2	5	1
2	5	3	1	4
5	1	4 >	2	3

36

3	x	9	+	2	=	29
x		+		x		
5	x	4	−	1	=	19
+		−		x		
6	x	8	+	7	=	55
=		=		=		
21		5		14		

38

6 – Reading along each row, deduct each number from the preceding number.

37

8	3	2	1		4	7		9	7	4
9	8	7	4		5	9		8	1	2
	9	3	2		3	2	6	4	1	
9	7	8	2	1	4		1	7		
8	9	6		3	6	9		4	2	3
4	1		5	4		1	6		1	7
	8	9	5			9	7	6	8	
3	9	4						1	3	9
5	8	9	6			6	3	5		
1	7		5	1		9	5		8	5
2	5	1		2	1	3		8	9	2
	7	9		6	7	2	4	3	1	
2	1	6	3	5		8	7	9		
8	6	9		3	1		3	7	9	1
4	3	8		8	5		1	6	8	2

39

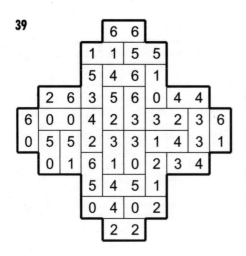

40

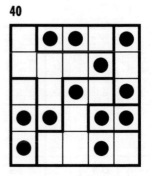

41

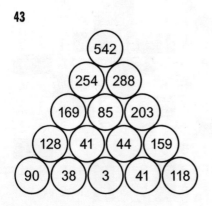

Clocks lose 50 minutes, gain 60 minutes, lose 70 minutes and gain 80 minutes.

42

Circle = 2, cross = 3, pentagon = 9, square = 4, star = 5.

43

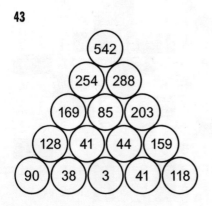

44

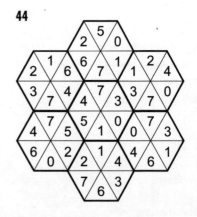

45

1		0	2	●		●	●	1	
●			●	3		3	3		
		2	●				1	●	●
0		2	●	3					
	1			●				0	
		●	3	2		0			
	●	3	2	●				●	2
2	●				●	3	●	5	●
	●				2	●	4	●	●
●	3	●	2					●	3

46

1	3	4	2	6	5
5	1	2	3	4	6
2	6	5	1	3	4
6	5	3	4	2	1
3	4	6	5	1	2
4	2	1	6	5	3

47

1	2	1	2	1	2	1	4
3	4	3	3	4	3	3	2
2	4	3	4	1	2	1	3
1	2	2	4	4	3	4	2
1	4	1	3	2	1	3	1
2	3	4	1	3	2	4	2
1	2	4	4	1	1	3	3
4	3	1	3	2	2	4	4

48

1	4	3	2	5
2	1	4	5	3
3	5	2	4	1
4	3	5	1	2
5	2	1	3	4

49

21	12	38	57	88	5
42	36	36	21	41	45
53	46	36	14	26	46
14	51	41	58	26	31
25	54	43	49	13	37
66	22	27	22	27	57

50

3	2	2	3	3	3
1	4	4	2	2	3
1	4	4	2	2	3
1	2	2	2	2	4
1	2	2	2	2	4
3	1	1	4	4	3

51

6	7	3	5	9	8	4	1	2
1	2	4	3	7	6	5	9	8
5	9	8	2	4	1	3	7	6
4	1	2	9	3	5	8	6	7
7	8	9	1	6	4	2	3	5
3	6	5	7	8	2	1	4	9
9	5	6	8	1	3	7	2	4
2	3	7	4	5	9	6	8	1
8	4	1	6	2	7	9	5	3

52

5	3	2 >	1	4
3	1	4	2	5
2 <	4	1	5 >	3
1	5	3	4 >	2
4	2	5	3	1

53

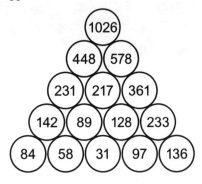

54

1	x	6	–	2	=	4
x		x		x		
3	x	5	–	4	=	11
+		+		x		
8	x	9	+	7	=	79
=		=		=		
11		39		56		

55

1	2	7		5	9	7			7	3	9
3	1	8		1	7	4	5	3	2	6	
5	7	9	8	2		5	3	2	1		
		5	2		2	9	7		8	9	
2	1	4		1	5	6	8	2	4	3	
7	3		9	2		8	9	6			
	2	5	4	3	1		2	3	4	1	
1	9	7		7	9	8		5	9	8	
4	8	9	7		5	9	2	1	6		
	6	1	2		5	1		7	4		
6	4	8	2	5	9	7		7	8	1	
7	2		3	1	5		1	3			
	6	9	8	7		2	7	5	1	4	
8	3	7	5	6	9	4		9	2	1	
6	1	8		3	7	1		8	3	9	

56

160 – Reading down each column, multiply the first number by 4, then deduct 4, then multiply by 5, then deduct 5, then multiply by 6, then deduct 6.

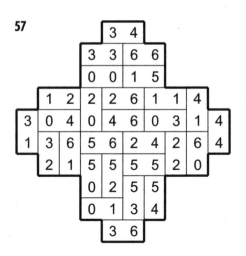

57

58

59

24	3	17	13	15	20	4	96
6	12	11	9	24	27	3	92
25	19	10	13	21	19	14	121
27	1	30	22	17	16	29	142
5	4	16	23	14	7	19	88
8	24	14	9	6	12	23	96
13	2	15	28	22	1	8	89

108	65	113	117	119	102	100	102

127

60

Clocks lose 1 hour 11 minutes each time.

61

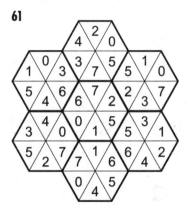

62
8788

63
1 x 11 + 21 − 31 = 1

149

64

0		●	3	●	●	●		●
	2	●		2	3	2		1
1				3		1		1
●	1		1	●	●	●		●
	2			5	●		●	●
	●	1			●	2		5
●			0				●	●
1		1		1	0		1	
		1	●	3				1
	0			●	●	●	2	●

65

2	6	5	4	3	1
6	3	1	5	2	4
3	2	4	6	1	5
4	1	6	2	5	3
5	4	3	1	6	2
1	5	2	3	4	6

66

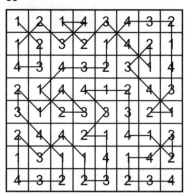

67

◇ 1	⬡ 4	★ 5	○ 3	□ 2
○ 2	□ 1	◇ 4	⬡ 5	★ 3
★ 4	○ 5	□ 3	◇ 2	⬡ 1
□ 5	◇ 3	⬡ 2	★ 1	○ 4
⬡ 3	★ 2	○ 1	□ 4	◇ 5

68

13	10	15	23	31	29
30	20	21	10	21	19
26	38	20	4	13	20
15	19	21	36	11	19
13	22	16	34	17	19
24	12	28	14	28	15

69

2	3	3	4	4	2
4	3	3	1	1	2
4	3	3	1	1	2
4	2	2	3	3	4
4	2	2	3	3	4
1	1	1	4	4	1

70

6	5	8	9	2	4	3	7	1
3	1	9	5	7	6	4	8	2
2	7	4	8	3	1	6	9	5
4	9	7	2	1	8	5	6	3
5	8	3	6	4	7	2	1	9
1	6	2	3	9	5	8	4	7
7	3	1	4	8	2	9	5	6
9	4	6	7	5	3	1	2	8
8	2	5	1	6	9	7	3	4

71

1	5	4	3	2
2	3	1 < 4	5
4 > 1	2	5	3
5	2	3	1	4
3	4	5	2	1

72

The value of the letter in the central square is equal to the difference between the sum total of the values in the top squares and the sum total of the values in the bottom squares. Thus the missing value is 17, so the missing letter is Q.

73

$6 \times 7 - 8 + 9 = 43$

74

9	x	7	x	2	=	126
x		−		+		
8	−	3	x	4	=	20
−		x		x		
1	x	5	x	6	=	30
=		=		=		
71		20		36		

76

339 – From the top left corner, follow a clockwise path around and spiral towards the centre, adding 7 to each number every time.

75

6	4		4	2	5	1		3	2	1
2	6	9	8	4	7	3		8	7	2
	5	7	3		9	5	8	7	6	4
8	2	1		7	8	4	6	9		
9	3	5	2	1		2	9		4	9
2	1		9	8	6		2	5	1	3
	8	1		5	9	7	8	3	6	
3	6	1		1	2	6		9	2	8
1	7	2	5	4	3		1	6		
2	8	4	7		1	2	3		1	4
8	9		8	9		3	8	7	6	9
	2	6	4	3	1		9	3	8	
8	1	4	9	7	2		8	3	2	
5	2	1		8	7	1	9	5	4	3
9	4	5		2	1	3	7		7	9

77

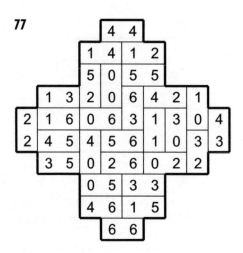

78

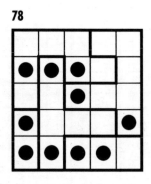

79

Clocks gain 2 hours 19 minutes,
2 hours 29 minutes, 2 hours 39
minutes and 2 hours 49 minutes.

81

2827

80

84

13	11	9	7	22	21	18	101
17	4	27	29	19	3	9	108
15	24	16	22	28	1	10	116
20	13	21	14	18	7	26	119
2	11	12	25	16	23	5	94
27	6	19	14	9	10	30	115
3	22	16	15	12	4	26	98
97	91	120	126	124	69	124	99

82

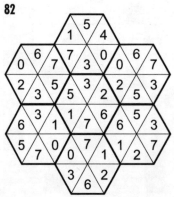

83

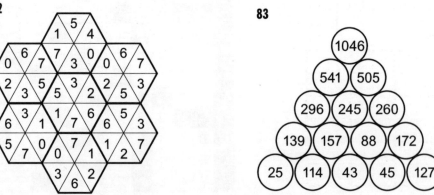

84

1	●	1				0		●	●
				0					3
3	●		1				0		●
●	●	3		●			0		●
	4	●	5		2	0			2
	3	●	●	●	1			3	●
	●		4				2	●	●
3	●	2	1	●				●	●
●	3		2		1		3	●	3
		●		0			●	2	

85

6	5	4	2	3	1
1	3	2	5	6	4
4	2	3	1	5	6
2	6	5	4	1	3
5	1	6	3	4	2
3	4	1	6	2	5

86

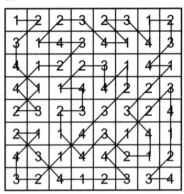

87

5	4	1	2	3
3	5	2	4	1
1	3	4	5	2
2	1	5	3	4
4	2	3	1	5

88

17	33	21	40	39	45
49	32	32	14	35	33
45	48	32	18	19	33
18	42	35	46	22	32
16	12	46	39	49	33
50	28	29	38	31	19

89

1	2	2	3	3	2
4	1	1	1	1	4
4	1	1	1	1	4
3	3	3	4	4	1
3	3	3	4	4	1
4	4	4	2	2	2

90

6	3	2	5	8	9	4	7	1
9	5	1	7	3	4	6	2	8
4	7	8	6	2	1	5	9	3
8	9	4	3	7	6	1	5	2
2	1	7	9	5	8	3	4	6
3	6	5	1	4	2	7	8	9
5	4	9	8	6	3	2	1	7
7	8	6	2	1	5	9	3	4
1	2	3	4	9	7	8	6	5

93

$9 \times 6 - 8 + 7 = 53$

94

2	x	7	x	6	=	84
−		x		x		
1	+	4	x	8	=	40
x		−		−		
9	x	3	+	5	=	32
=		=		=		
9		25		43		

96

0 – In the first row, add 4 to the first number, subtract 7 from the second, add 4 to the third, etc; in the second row, add 7 to the first number, subtract 4 from the second, etc; then repeat this process for the remaining rows, adding and subtracting 4 and 7 alternately.

91

2	1	5	3	4
4	5	3	2	1
3	4	2	1	5
5	3	1	4	2
1	2	4	5	3

92

The value of the letter in the top right is deducted from that in the top left, and the value of the letter in the bottom right is deducted from that in the bottom left, then the result of the bottom sum is taken from that of the top to give the value in the central square. Thus the missing value is 4, so the missing letter is D.

95

		8			
				1	5
6		5			
				5	9
	3				
9		6		6	4

97

Circle = 4, cross = 9, pentagon = 2, square = 3, star = 8.

98

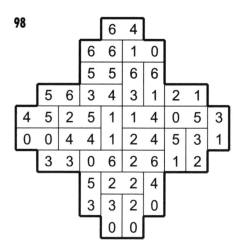

99

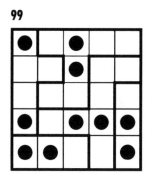

100

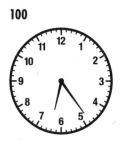

Clocks alternately lose 85 minutes and gain 143 minutes each time.

101

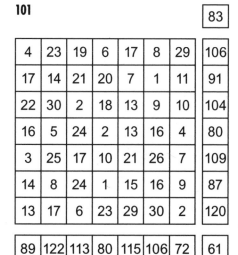

							83
4	23	19	6	17	8	29	106
17	14	21	20	7	1	11	91
22	30	2	18	13	9	10	104
16	5	24	2	13	16	4	80
3	25	17	10	21	26	7	109
14	8	24	1	15	16	9	87
13	17	6	23	29	30	2	120
89	122	113	80	115	106	72	61

102

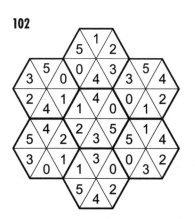

103

2955

104

			●	1			●	2	
	1	2			1	2	●		2
	●		3	●	2	1	3	●	●
1	1		●	●				●	●
1				3	●		3	3	3
●		1		3		●		●	2
●		2	●	3	●	3	●	3	●
	2		●	3	1				
0		●	2	1			●	2	
						●	3	●	

105

3	1	4	5	2	6
2	3	1	4	6	5
6	2	3	1	5	4
5	6	2	3	4	1
1	4	5	6	3	2
4	5	6	2	1	3

106

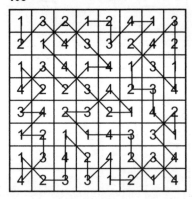

107

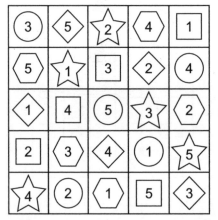

108

29	27	24	21	19	25
25	24	20	18	25	33
19	29	24	34	20	19
22	31	25	14	17	36
22	8	26	33	39	17
28	26	26	25	25	15

109

1	1	1	2	2	4
2	3	3	4	4	4
2	3	3	4	4	4
1	3	3	2	2	2
1	3	3	2	2	2
2	1	1	4	4	3

110

5	7	8	9	2	1	6	3	4
6	3	1	5	8	4	7	2	9
4	9	2	6	3	7	8	1	5
1	2	6	4	7	5	3	9	8
9	4	3	8	6	2	5	7	1
8	5	7	3	1	9	2	4	6
7	8	9	2	4	6	1	5	3
3	1	4	7	5	8	9	6	2
2	6	5	1	9	3	4	8	7

111

3 >	2	4	5	1
1	3	2	4	5
2	5	3	1	4
5	4	1 <	2	3
4	1	5	3	2

112

6	+	9	−	7	=	8
−	■	+	■	x		
4	−	3	x	8	=	8
+	■	x	■	+		
2	x	1	x	5	=	10
=		=		=		
4		12		61		

113

9		6		6			
						4	
			2				
	4						
				6		3	
		4					
1				9		7	

114

113 – In the first column, deduct 11 then 12 from each successive number; in the second column, deduct 12 then 13; in the third, deduct 13 then 14; in the fourth, deduct 14 then 15; in the fifth, deduct 15 then 16; in the sixth, deduct 16 then 17; and in the seventh, deduct 17 then 18.

115

Circle = 5, cross = 7, pentagon = 3, square = 4, star = 8.

116

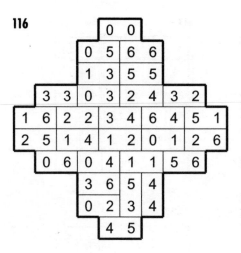

			0	0					
		0	5	6	6				
		1	3	5	5				
	3	3	0	3	2	4	3	2	
1	6	2	2	3	4	6	4	5	1
2	5	1	4	1	2	0	1	2	6
	0	6	0	4	1	1	5	6	
		3	6	5	4				
		0	2	3	4				
			4	5					

117

118

The clock moves back 4 hours 41 minutes, forwards 1 hour 44 minutes, back 4 hours 41 minutes and forwards 1 hour 44 minutes.

119

							98
20	15	6	17	8	14	7	87
24	16	9	4	1	26	5	85
3	18	30	2	11	12	13	89
27	6	13	19	21	20	2	108
10	14	8	22	29	24	9	116
14	12	6	27	19	10	16	104
15	20	21	17	4	11	12	100
113	101	93	108	93	117	64	136

120

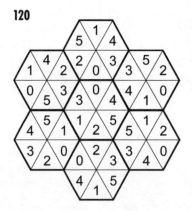

121

1210

122

2		1		●	●	3	●	●	●
●	●	2	2	●	●				3
3	●			3	3	2		●	1
2		2		1	●				2
●		●	2		2			1	●
	●		2		●	1		1	
	2		1			2		1	
●	2		2		●	2			●
	2	●	●	4	●		1		
	1		●			1	2	●	1

123

5	1	4	3	2	6
6	5	1	4	3	2
2	4	5	1	6	3
1	3	2	6	5	4
4	6	3	2	1	5
3	2	6	5	4	1

124

1	3	4	1	2	1	4	3
2	4	2	3	3	4	2	2
3	4	1	3	2	3	1	1
2	1	2	4	1	3	4	2
2	1	3	1	4	4	1	3
3	4	2	3	2	1	4	2
4	3	2	1	4	3	1	3
1	2	4	1	4	3	2	4

125

(4) circle	⟨1⟩ diamond	[2] square	⬡3 hexagon	☆5 star
⟨2⟩ hexagon	☆3 star	(5) circle	⟨4⟩ diamond	[1] square
⟨5⟩ diamond	[4] square	⬡1 hexagon	☆2 star	(3) circle
[3] square	⬡5 hexagon	☆4 star	(1) circle	⟨2⟩ diamond
☆1 star	(2) circle	⟨3⟩ diamond	[5] square	⬡4 hexagon

126

53	15	23	28	38	39
40	32	29	23	36	36
31	52	32	34	15	32
31	52	36	30	13	34
22	32	34	46	28	34
19	13	42	35	66	21

127

4	1	1	3	3	2
2	4	4	1	1	1
2	4	4	1	1	1
3	4	4	1	1	2
3	4	4	1	1	2
1	2	2	2	2	3

128

1	4	7	8	3	5	2	9	6
3	5	9	6	4	2	1	7	8
8	2	6	1	9	7	3	4	5
4	9	8	5	1	6	7	2	3
2	1	5	3	7	8	4	6	9
7	6	3	9	2	4	8	5	1
5	7	1	2	8	9	6	3	4
6	3	4	7	5	1	9	8	2
9	8	2	4	6	3	5	1	7

129

2	3	4	5	1
5	4	3	1	2
3	2	1	4	5
4	1	5	2	3
1	5	2	3	4

130

3	+	4	−	2	=	5
x			−		x	
9	+	1	+	6	=	16
x			x		−	
5	x	7	−	8	=	27
=		=		=		
135		21		4		

131

6		1			8		
1		7		8			
							1
1		8					
							2
2			9				

132

5 – In the first column, add 4 to the first number, subtract 6 from the second, add 4 to the third, etc; in the second column, add 6 to the first number, subtract 4 from the second, etc; then repeat this process for the remaining columns, adding and subtracting 4 and/or 6 alternately.

133

Circle = 2, cross = 4, pentagon = 6, square = 1, star = 8.